I0463485

Arthur Jackson CTC
CPR TRAINING THAT WORKS

Life Skills Training Manual

Arthur Jackson CTC, Inc.

Purpose of this guide

This training guide, Version 4.0, is solely intended to facilitate the training of Lay Responders (persons without specialized medical training) in the use of Basic Life Skills to stabilize injured individuals in accordance with established National and International Standards.

Certifications for Arthur Jackson CTC Inc. may only be issued by an Arthur Jackson CTC authorized instructor when a student has successfully completed the required knowledge and demonstrated the required competency.

Life Skills Training Manual / Arthur Jackson CTC Inc.
ISBN 978-1-329-44849-0

Contents

Introduction to Emergencies for Lay Responders

The Lay Responder

A Lay Responder is a person without specialized medical training who can provide initial emergency procedures using limited equipment to stabilize an injured person until Emergency Medical personnel can arrive. The definition(s) of a lay responder are as follows:

- Layperson (n.)

    1. One who is not a cleric.

    2. One who is a nonprofessional in a given field.

- Responder (n.)

    1. A person or thing that responds.

    2. The part of a transponder that transmits the reply.

American Heritage® Dictionary of the English Language, Fifth Edition. Copyright © 2016 by Houghton Mifflin Harcourt Publishing Company. Published by Houghton Mifflin Harcourt Publishing Company. All right reserved.

THE GOAL OF THE LAY RESPONDER IS TO:

- Recognize the presence of a medical emergency
- Make a decision to help
- Make the appropriate checks for safety (personal and for the injured)
- Activate the EMS system
- Provide care in accordance with established law and standards

Barriers to Act

Providing care to the injured comes with responsibilities and concerns. In general, three major concerns have been identified as the leading cause of fear and the reason there are barriers to act among the public. The American Heart Association reported that these concerns result in 70% of injured persons not receiving prompt care in an emergency. Every person will bring a different level of fear and concern into an emergency. The key to overcoming these barriers is to fully understand the rights and responsibilities of the lay responder. The three areas of concern are:

- Legal considerations – fear of being sued
- Disease – fear of communicable disease transmission
- Lack of training – fear of not knowing what to do

Each area brings with it special fears for a lay person providing aid to an injured individual in an emergency. These fears are not unreasonable and should not be considered as weaknesses. They are the normal reaction of many persons and can be dealt with effectively by providing proper information and training. Each concern will be evaluated in this section to provide lay persons with proper information and the understanding that if established procedures are followed, there is no need to be afraid.

Legal Considerations

Legal considerations can be broken into two areas: international laws and national laws. Both have significant considerations for lay persons working in an emergency. In the age of mobility and travel among individuals and groups, it is important to understand the duties imposed on lay people while present in other countries.

International Laws

Today most countries have laws regulating the legal responsibilities of lay individuals to the injured in an emergency. These laws are a response to the establishment of the International Liaison Committee on Resuscitation by the American Heart Association and other international agencies in the 1990s. This organization establishes the international standards for CPR and emergency care. These standards are the basis for growth in international laws governing CPR and emergency treatment.

INTERNATIONAL LIAISON COMMITTEE ON RESUSCITATION (ILCOR)

ILCOR was formed in 1992 to provide a forum for connection between principal resuscitation organizations worldwide. At present, ILCOR comprises representatives of:

- American Heart Association (AHA)
- European Resuscitation Council (ERC)
- Heart and Stroke Foundation of Canada (HSFC)
- Australian and New Zealand Committee on Resuscitation (ANZCOR)
- Resuscitation Councils of Southern Africa (RCSA)
- Inter American Heart Foundation (IAHF)
- Resuscitation Council of Asia (RCA)

The objectives of the ILCOR are to:

- Provide a forum for discussion and for coordination of all aspects of cardiopulmonary and cerebral resuscitation worldwide.
- Foster scientific research in areas of resuscitation where there is a lack of data or where there is controversy.
- Disseminate information on training and education in resuscitation.
- Provide a mechanism for collecting, reviewing and sharing international scientific data on resuscitation.
- Produce statements on specific issues related to resuscitation that reflect international consensus.

ILCOR meets twice annually, once in the United States and once abroad. Every five (5) years, a global meeting is held to develop an international consensus document.

Taken From: ILCOR website

National Laws

Many countries will have in place laws to protect lay persons. These laws will differ from country to country and regions within a country, and you should become familiar with the laws of the country you are in, traveling through, or working in to understand how these laws will apply to you. In the United States, OSHA regulation and Federal Law are applicable to lay individuals

OSHA AND FEDERAL LAWS

a) On November 20, 2003, the American Heart Association and OSHA entered into an Alliance agreement to promote CPR. This agreement has been renewed every two years.

b) 1910.151(b) states, "In the absence of an infirmary clinic, or hospital in near proximity to the workplace which is used for the treatment of all injured employees, a person or persons shall be adequately trained to render first aid. Adequate first aid supplies should be readily available."

c) Although it is not an OSHA requirement that employers provide Cardiopulmonary Resuscitation (CPR) training, OSHA's "Guidelines for First Aid Training Programs" recommend that CPR training be a general element of a first aid program. It is recommended that employees receive refresher training to retain their knowledge of first aid procedures. Some occupations are now required to have CPR training every year:

- 1910.146 Permit-required Confined Spaces,
- 1910.266 Appendix B: Logging Operations – First-Aid and CPR Training
- 1910.269 Electric Power Generation, Transmission, and Distribution
- 1910.410 Qualifications of Dive Team
- 1926.950 Construction Subpart V Power Transmission and Distribution

d) Volunteer Protection Act of 1997 – Protection of volunteers acting with non-profits or government agencies from common negligence as long as they are acting accordance with their responsibilities to the organization.

Taken From: Osha.gov and 111 STAT. 218 PUBLIC LAW 105

State Laws

All states in the United States have passed laws known as Good Samaritan Laws to help encourage lay persons to aid in an emergency. These laws provide protection to lay people provided they meet the following criteria:

- Voluntarily assists (does not ask or seek any remuneration for providing care)
- Familiarity with the laws that govern your state
- Stays within the scope of their training
- Provides training in accordance with applicable law
- Asks permission to provide care (everyone has the right to refuse care)

It is important to familiarize yourself with the Good Samaritan Laws that govern your state since they can vary. In general, to receive protection, follow the "S.M.A.C.K." process:

S.M.A.C.K RULES TO FOLLOW

- **S** – Stay within the scope of your training
  - o Examples: no surgery and no administration of medicine
- **M** – No money or compensation for your care; only move a person if the scene is unsafe, necessary to give appropriate care, or necessary to reach seriously injured persons
- **A** – Ask permission
- **C** – Check for injuries and call 9-1-1
- **K** – Keep giving aid until relieved

7

Disease

The transmission of disease during an emergency is a top concern. During emergencies, body fluids present from the emergency can be infectious. However, following standard precautions will provide maximum protection from disease.

STANDARD PRECAUTIONS

These procedures are required for a basic level of infection control and are recommended for dealing with persons in an emergency. They include:

- Hygienic practices, particularly washing and drying hands before and after contact
- Use of protective barriers when necessary which may include gloves, gowns, plastic aprons, masks, eye shields or goggles as appropriate
- Appropriate handling and disposal of sharps and other contaminated or clinical waste
- Appropriate reprocessing of reusable equipment and instruments
- Use of proper cleansing techniques
- Use of environmental controls
- The implementation of standard precautions minimizes the risk of cross-infection from responder to patient, from patient to responder, and from person to person, even in high-risk situations. They are recommended for use with all persons, regardless of their perceived infectious status.

If a first aid kit and appropriate protective equipment is not available, the responder can use improvised barriers from material at hand such as trash can liners, clothing, or food wrappers. When using improvised barriers be alert to possible allergic reactions. These standard precautions require that each person use proper hand hygiene before, during and after each emergency. Proper cleaning of hands will prevent the spread of disease.

HAND HYGIENE

- Good hand hygiene including use of alcohol-based hand rubs (ABHR) and hand washing with soap and water is critical to reduce the risk of spreading infections in ambulatory care settings. It is recommended by the CDC and the World Health Organization (WHO) because of its activity against a broad spectrum of epidemiologically important pathogens.
- Key situations where hand hygiene should be performed include:
  - Before touching a patient, even if gloves will be worn
  - After contact with blood, body fluids, excretions or wound dressings
  - If hands will be moving from a contaminated-body site to a clean-body site
  - After glove removal
- Use soap and water when hands are visibly soiled (e.g., blood, body fluids), or after caring for patients with known or suspected infectious diarrhea (e.g., Clostridium difficile, norovirus). Otherwise, the preferred method of hand decontamination is with an alcohol-based hand rub.

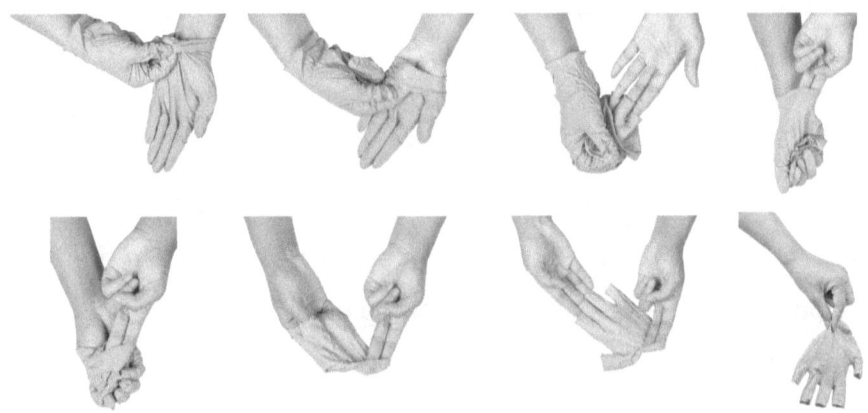

Proper hand hygiene includes the proper removal of contaminated gloves.

Lack of Training

The fear of "not knowing what to do" is the easiest to be dealt with by the responder. It only requires training in classes such as this. In these classes the three "C" method is used to keep the responder safe. Always do the following before entering any emergency:

1. **CHECK** the scene – Checking the scene is a three step process:
   a. Is it safe? Look, listen and smell for any danger
   b. Who do I go to first? Prioritize the injured
   c. Is there anyone that can help you?
2. **CALL** – Call 9-1-1. Calling is the most important action you can do to help someone who is ill or injured. This will dispatch the emergency medical response team insuring they arrive at the scene as quickly as possible. This is important! If you don't call 9-1-1, no emergency help will come.
   a. Who Should Call 9-1-1?
   b. If **SOMEONE** else is around, instruct **THEM** to call 9-1-1 so you can begin **CARE**.
   c. If you are **ALONE**: CALL 9-1-1 FIRST!
3. **CARE** – start to give care and continue giving care until relieved by another lay responder, an EMT, or Doctor.

Structure of the Emergency Medical System

The Emergency Medical System (EMS) is primarily responsible for the treatment of all injured persons. The lay responder provides initial assessment and basic treatment of injuries to stabilize the injured until EMS personnel can arrive. The EMS system is a hierarchal system of lay responders, responders, EMTs, Nurses, and Doctors.

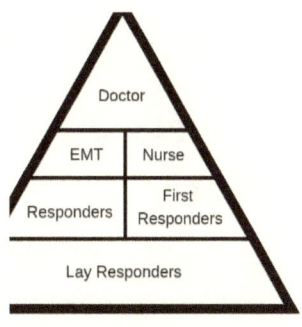

Lay responders are in many cases the first on the scene of an emergency and may assume the responsibility of providing care, calling 9-1-1 to alert EMS personnel of the emergency, and continuing to provide care to stabilize and maintain the injured until EMS can arrive.

Taking Personal History Information

When contacting the EMS system, you will ideally speak to one person, the dispatcher. Dispatchers are highly trained members of the EMS system and will begin the process of collecting valuable information about the injured. After ascertaining the location, type and number of injured, the questions will often be in the form of the acronym S.A.M.P.L.E.

THE SAMPLE QUESTIONS FOR PERSONAL HISTORY

- **S** for Symptoms: pain, nausea, lightheadedness, other things you cannot see
- **A** for Allergies: any known allergic reactions? What happens?
- **M** for Medications: anything legal or illegal? Why? How much?
- **P** for Pertinent Medical History: has anything like this happen before? Currently under a physician's care for anything?
- **L** for Last Intake and Output: when was food or drink last taken? How much? When were the most recent urination and defecation? Were they normal?
- **E** for Events: what led up to the accident or illness? Why did it happen?

Provide the dispatcher with all the pertinent information you have collected on the injured. At this moment the lay responder is in the same position as someone treating an injured person in a remote or rural location. They must collect information as accurately as possible and relay that information to the EMS personnel.

SOAP: THE WRITTEN REPORT AND VERBAL REPORT

In an emergency your brain tends to become a sieve instead of a bowl. The acronym SOAP reminds you to write everything down (to collect documentation) as soon as possible, as long as taking notes does not interfere with patient care. Retention of information for medical and legal reasons is important.

- **S** for Subjective Summary: a summary for who the patient is (including age and sex), what the patient is complaining about, and what happened to the patient.
- **O** for Objective/Observations: observations and results of patient exam, vital signs, and SAMPLE history.
- **A** for Assessment: what do you think is wrong?
- **P** for Plan: what you are going to do immediately for the patient? Answer the evacuation question (stay or go, fast or slow).

Priority of Injured

One of the most important things the lay responder will do is to prioritize the injured so that the person needing care the most will receive it promptly.

Lay responders do not diagnose injuries and require a simplified approach to triage (sorting injured people into groups based on their need). In any emergency, these are the injuries to attend to first:

- Unconscious – there are two critical reasons for people to be unconscious in an emergency
    1. Cardiac emergency – in this case it is important to begin CPR immediately to keep oxygenated blood flowing throughout the body.
    2. Blood loss – severe blood loss can cause a person to become unconscious in less than a minute, and to be beyond revival (dead) in less than two minutes.
- Bleeding – bleeding must be controlled, or it can lead to unconsciousness and death
- All other injuries – everything else is then treated

Cardiac System

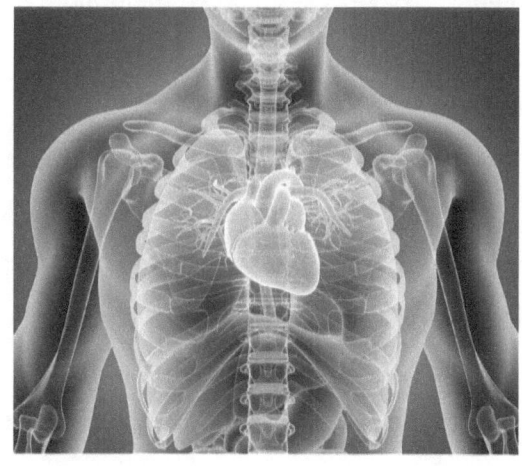

The cardiovascular system has three major functions: transportation of materials, protection from pathogens, and regulation of the body's homeostasis. Transportation: The cardiovascular system transports blood to almost all of the body's tissues. The blood delivers essential nutrients and oxygen and removes wastes and carbon dioxide to be processed or removed from the body. Hormones are transported throughout the body via the blood's liquid plasma.

Protection: The cardiovascular system protects the body through its white blood cells. White blood cells clean up cellular debris and fight pathogens that have entered the body. Platelets and red blood cells form scabs to seal wounds and prevent pathogens from entering the body and liquids from leaking out. Blood also carries antibodies that provide specific immunity to pathogens that the body has previously been exposed to or has been vaccinated against.

Regulation: The cardiovascular system is instrumental in the body's ability to maintain homeostatic control of several internal conditions. Blood vessels help maintain a stable body temperature by controlling the blood flow to the surface of the skin. Blood vessels near the skin's surface open during times of overheating to allow hot blood to dump its heat into the body's surroundings. In the case of hypothermia, these blood vessels constrict to keep blood flowing only to vital organs in the body's core. Blood also helps balance the body's pH due to the presence of bicarbonate ions, which act as a buffer solution. Finally, the albumins in blood plasma help to balance the osmotic concentration of the body's cells by maintaining an isotonic environment.

The Circulatory Pump

The heart is a four-chambered "double pump" where each side (left and right) operates as a separate pump. The left and right sides of the heart are separated by a muscular wall of tissue known as the septum of the heart. The right side of the heart receives deoxygenated blood from the systemic veins and pumps it to the lungs for oxygenation. The left side of the heart receives oxygenated blood from the lungs and pumps it through the systemic arteries to the tissues of the body. Each heartbeat results in the simultaneous pumping of both sides of the heart, making the heart a very efficient pump.

Types of Cardiac Emergencies

Cardiac emergencies can occur without warning to anyone at any time. Cardiac emergencies are a leading cause of death among adults throughout the world. The two leading causes of cardiac emergencies are:

- Arrhythmia – the term arrhythmia refers to any change from the normal sequence of electrical impulses. The electrical impulses may happen too fast, too slowly or erratically, causing the heart to beat too fast, too slowly or erratically. When the heart doesn't beat properly, it can't pump blood effectively. When the heart doesn't pump blood effectively, the lungs, brain and all other organs can't work properly and may shut down or be damaged.

- Heart Attack – a heart attack occurs when the blood flow that brings oxygen to the heart muscle is severely reduced or cut off completely. This happens because coronary arteries that supply the heart muscle with blood flow can slowly become narrow from a buildup of fat, cholesterol and other substances that together are called plaque. This slow process is known as atherosclerosis. When a plaque in a heart artery breaks, a blood clot forms around the plaque. This blood clot can block the blood flow through the heart muscle. When the heart muscle is starved for oxygen and nutrients, it can become damaged and the result is cardiac arrest.

Cardiac arrest is a life-threatening condition and requires swift action to preserve oxygen flow to vital organs.

Cardiac Chain of Care

The Cardiac Chain of Care is the model used to describe the best approach of treating cardiac arrest. This model is based on the recommendations, research and experience of medical professionals worldwide.

THE CARDIAC CHAIN OF CARE

1. Early Recognition
2. Early CPR
3. Early use of an AED
4. Early Access to Advanced Medical Care
5. Access to Appropriate After Care

The first link to the emergency response system, **Early Recognition**, includes early recognition of the cardiac emergency and early notification of rescue personnel via a universal 9-1-1 telephone system (or other emergency number) as well as an internal alert system within specific facilities to trigger a response by designated trained and equipped personnel. The second link, **Early CPR**, is a set of actions that the rescuer performs in sequence to assess and support airway, breathing and circulation.

The third link, **Early use of an AED**, is the delivery of a shock to the heart to convert the heart's rhythm from Ventricular Fibrillation back to a normal heart rhythm.

The fourth link, **Access to Advanced Medical Care**, relates to the response of highly trained and equipped pre-hospital EMS personnel (paramedics) who can respond to the patient and provide for the administration of drugs, advanced airway procedures, and other interventions and protocols prior to the arrival of the patient at an advanced care facility.

However, in order for the patient to have the best chance of surviving an out-of-hospital cardiac arrest, CPR and early defibrillation must be provided within the first 4 minutes of the cardiac arrest (the American Heart Association recommends 3 minutes) followed by Advanced Life Support within the first 8 minutes of the arrest.

Cardio Pulmonary Resuscitation (CPR)

The primary method of dealing with a cardiac emergency is CPR. If an AED is present, it should be used as soon as it is available. However, AEDs will not be effective in cases of heart attack. A heart attack is the result of a damaged heart muscle. The AED will direct you to do CPR, as shocks to the system will not remedy the problem. CPR is the primary method of treatment for all cardiac emergencies.

CPR will restore some blood flow and oxygen to critical organs. CPR as we know it today has been around for only a few decades.

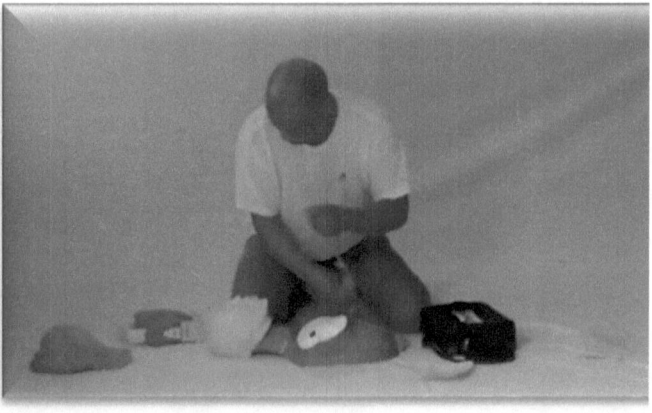

BRIEF HISTORY OF CPR

- 1956 – Peter Safer and James Elam invented mouth-to-mouth resuscitation
- 1960 – W.B. Kouwenhoven, J.R. Jude and G.G. Knickerbocker began to use what was termed Cardiopulmonary Resuscitation or CPR
- 1965 – Modern CPR established
- 1992 – ILCOR established (International Liaison Committee on Resuscitation)
- 2005 – First international consensus document by ILCOR
- 2010 – Current consensus document by ILCOR

In 2005, with the results of ILCOR and work in the European community, we finally understood that the method of thirty compressions and two rescue breaths was the most effective method of providing CPR. As a result of the global consensus, thirty compressions and two rescue breaths as a cycle of CPR was made a global standard.

WHY RATIO OF 30:2

- The European Resuscitation Council in conjunction with the American Heart Association and other relevant worldwide bodies issued new resuscitation procedures on November 28, 2005.

- They were released after collective concern at the lack of skill retention by individuals who had completed first aid courses. Additionally, there were many different protocols of CPR, depending on injury/non-injury/child CPR/2-person CPR. There was also some doubt raised by clinicians as to the clinical benefit of the old system (not that it was ineffective but that there may be a more efficient technique).

- Hence the latest guidelines, the most striking change being 30 chest compressions where before we carried out 15 compressions.

- WHY? It was noted that under the old regime of 15:2, that it could take up to 10 compressions to raise the pressure inside the chest cavity (intro-thoracic pressure) to the point where oxygen perfusion would take place into the heart muscle. This meant that only the last 5 compressions were of clinical benefit to the heart. After the 15 compressions we would give 2 rescue breaths, during which time the intra-thoracic pressure would naturally drop, and we would have to repeat the whole process.

- With the above in mind, it was decided that instead of 15 compressions we would carry out 30 compressions. That way we could be confident that at least 20 of the 30 compressions was perfusing oxygen into the heart muscle, thereby giving the casualty the best possible chance of survival
- To address the skill retention issue, this change was adopted over the entire age range from infant up to adults. CPR suddenly got easier to understand and remember!

https://medrocktraining.co.uk/why-30-compressions/

Correct application of the chest compressions is essential for effective CPR. Here's advice from the American Heart Association:

- **Untrained.** If you're not trained in CPR, then provide hands-only CPR. That means uninterrupted chest compressions of about 100 a minute until paramedics arrive (described in more detail below). You don't need to try rescue breathing.

- **Trained and ready to go.** If you're well-trained and confident in your ability, begin with chest compressions instead of first checking the airway and doing rescue breathing. Start CPR with 30 chest compressions before checking the airway and giving rescue breaths.

- **Trained but rusty.** If you've previously received CPR training but you're not confident in your abilities, then just do chest compressions at a rate of about 100 a minute.

(http://www.mayoclinic.org/first-aid/first-aid-cpr/basics/art-20056600)

The reason untrained individuals and those individuals with rusty skills are urged to do just compressions is to avoid gastric distension.

GASTRIC DISTENTION

- During rescue breathing or CPR, air may enter the casualty's esophagus (the tube leading from the throat to the stomach) and cause the stomach to inflate. This condition is called gastric distention. Gastric distention can be caused by:
  - o The rescuer delivering the ventilations with too much force
  - o Improperly positioning the casualty's head (airway not open)
  - o An obstruction in the casualty's airway preventing his lungs from filling quickly.
- Gastric distention can cause vomiting and may decrease the lung volume by pushing up on the diaphragm
- (1) if the stomach becomes distended, reposition the casualty's airway. Watch for the rise and fall of the casualty's chest and breathe only hard enough to cause the chest to fully rise. Continue administering rescue breathing. Do not push on the casualty's abdomen in an attempt to reduce the distention since the pressure could cause the casualty to vomit.

- (2) Gastric distention can be corrected by adjusting the airway and adjusting the force of the ventilations. More advanced procedures may be appropriate, such as the use of a nasogastric tube to decompress the abdomen. If this is not available, then proper ventilation and opening of the airway are the primary methods to reduce further distention.

http://nursing411.org/Courses/MD0532_Cardiopulmonary_Resuscitation/4-04_Cardiopulmonary_Resuscitation.html

Differing Symptoms Between Women and Men

Women may exhibit different signals to Cardiac Arrest than men.

- Pain in arm(s), back, neck, or jaw
  - This type of pain is more common in women than in men. The pain can be gradual or sudden, and it may wax and wane before becoming intense. If the person is asleep, it may wake him/her up.
- Stomach pain
  - Sometimes people mistake heart attack stomach pain with heartburn, the flu, or a stomach ulcer. Other times, a woman might experience severe abdominal pressure that feels like an elephant sitting on her stomach.
- Shortness of breath, nausea, or lightheadedness
  - Trouble breathing for no apparent reason could be a heart attack, especially if the person has one or more other symptoms.
- Fatigue
  - Some women who have heart attacks feel extremely tired, even if they have been sitting still for a while or haven't moved much.

http://www.webmd.com/heart-disease/features/womens-heart-attack-symptoms

When Should You Stop CPR?

- Scene becomes unsafe
- Relieved by another responder, EMT, or physician
- AED arrives on the scene
- Victim shows signs of life
- You are too exhausted to continue

Chest Compressions

Be sure the person is face up on his/her back on a hard, flat surface. Kneel next to chest.

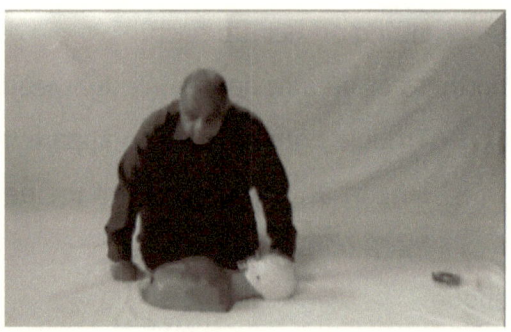

Place heel of one hand in the center of the chest.

Place the other hand on top of the first, with your shoulders directly above your hands. Stiffen elbows and use your upper body weight to compress chest.

Compressions are straight down at least two inches. Lift hands between compressions to allow chest to expand. Rate of compressions is about 100 per minute.

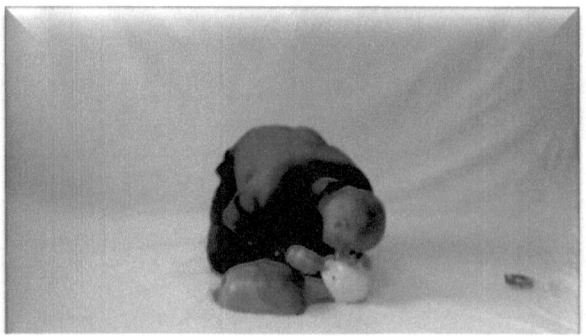

Rescue breaths applied after thirty compressions. Be sure to tilt head back to open airway and that airway is clear. Rescue breath is provided at normal breathing pressure.

Checking the Unresponsive Adult

When you first find someone that you suspect is unconscious,
follow these steps:

- Put gloves on and use your senses to check the scene for
  safety.
- Check the person for consciousness by tapping their
  shoulder and shouting "Are you okay? Are you okay?"
- If there is no response, tell someone "Call 9-1-1, the
  person is unconscious, and bring the on-site AED."
- If the person is facing down, roll the person as a unit onto
  his/her back.
  - Position yourself so that you are facing the back of the
    person's head
  - Move the arm that is closest to you up next to the
    head
  - Place one of your hands at the back of the neck and
    your other on the hip
  - Gently roll the person towards you keeping the head,
    neck and back in a straight line
- When the person is on their back, check their breathing.
  - Open the airway by placing the palm of one hand
    on the forehead and two fingers on the other hand
    under the bony part of the chin
  - Carefully tilt the head back as you lift up on the
    chin

- Keep the head tilted back and chin lifted to maintain an open airway
- Lean down and put your ear close to the person's mouth and nose and check for "signals of life"
- Look at the chest to see if it is clearly rising and falling
- Listen, and feel for movement and normal breathing for no more than 10 seconds. Count out loud, "one one-thousand, two one-thousand, three one-thousand..."
- If they are not breathing, place a breathing barrier on their mouth and give two rescue breaths ("breath, breath")
- After you check for "signs of life," quickly scan the entire body for sever bleeding as you get into position to perform CPR or use the AED

• If the person shows "signs of life," put in "recovery" position, keep airway open, monitor ABC's and provide care for the conditions found until the EMS arrives.

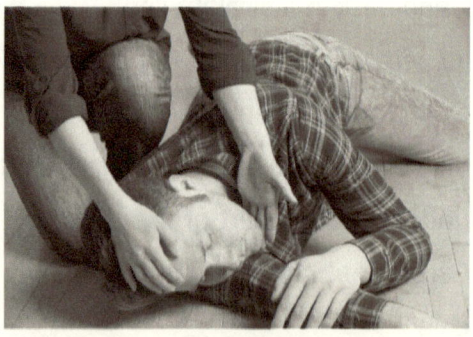

The Recovery Position

Checking the Conscious Adult

When the individual is conscious and alert, you should make an initial assessment of the individual and collect information for the EMS professionals when they arrive.

COLLECT INFORMATION FROM THE INJURED PERSON USING THE LETTERS OF THE WORD "SAMPLE":

- S – Symptoms such as pain, nausea, light headedness
- A – Allergies (known allergic reactions and what happens)
- M – medications they are taking
- P – prior pertinent medical conditions
- L – last time they had intake of food or drink
- E – events that led them to be injured or ill

This information should be collected from the injured person and made available to the EMS personnel when they arrive on scene. This information will help in decision making about the individual's medication treatment, preparation for surgery, or other advanced care.

The process of talking to the injured can have a calming effect, and the information obtained will be useful to EMS personnel when they arrive on the scene and to hospital personnel after the person arrives there.

Automatic External Defibrillator (AED)

The AED is a life saving device designed to analyze and treat cardiac emergencies. The AED has been in existence for a very short period of time.

TIME LINE FOR AED

- 1930 – William B. Kouenhoven of Johns Hopkins Hospital develops closed chest defibrillation using alternating current.
- 1947 – Claude Beck, Professor at Case Western Reserve University, successfully resuscitates a fourteen-year-old boy.
- 1960 – Frank Pantridge develops the first portable defibrillator.
- 2002 – First home use of an Automatic External defibrillator is approved.

Using an AED is a simple process: open the unit or turn on the power and follow the verbal instruction given by the device.

USING AN AUTOMATED EXTERNAL DEFIBRILLATOR:

- AEDs are user-friendly devices that untrained bystanders can use to save the life of someone having SCA.
  - Before using an AED, check for puddles or water near the unconscious person. Move him or her to a dry area and stay away from wetness when delivering shocks (water conducts electricity).
  - Turn on the AED's power. The device will give you step-by-step instructions. You'll hear voice prompts and see prompts on a screen.
  - Expose the person's chest. If the person's chest is wet, dry it. AEDs have sticky pads with sensors called electrodes. Apply the pads to the person's chest as pictured on the AED's instructions.
  - Place one pad on the right center of the person's chest above the nipple. Place the other pad slightly below the other nipple and to the left of the ribcage.

AUTOMATED EXTERNAL DEFIBRILLATOR:

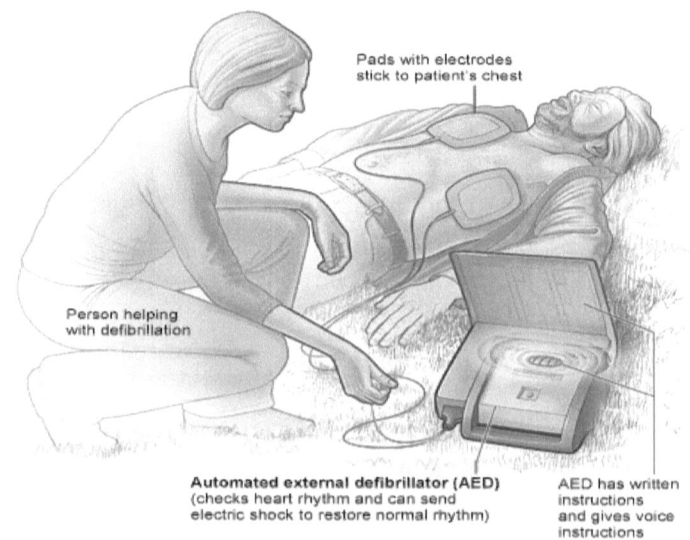

Pads with electrodes
stick to patient's chest

Person helping
with defibrillation

Automated external defibrillator (AED)
(checks heart rhythm and can send
electric shock to restore normal rhythm)

AED has written
instructions
and gives voice
instructions

The image shows a typical setup using an automated external defibrillator (AED). The AED has step-by-step instructions and voice prompts that enable an untrained bystander to correctly use the machine.

Make sure the sticky pads have good connection with the skin. If the connection isn't good, the machine may repeat the phrase "check electrodes." If the person has a lot of chest hair, you may have to trim it. (AEDs usually come with a kit that includes scissors and/or a razor.) If the person is wearing a medication patch that's in the way, remove it and clean the medicine from the skin before applying the sticky pads.

Remove metal necklaces and underwire bras. The metal may conduct electricity and cause burns. You can cut the center of the bra and pull it away from the skin.

Check the person for implanted medical devices, such as a pacemaker or implantable cardioverter defibrillator. (The outline of these devices is visible under the skin on the chest or abdomen, and the person may be wearing a medical alert bracelet.) Also check for body piercings.

Move the defibrillator pads at least 1 inch away from implanted devices or piercings so the electric current can flow freely between the pads.

Check that the wires from the electrodes are connected to the AED. Make sure no one is touching the person, and then press the AED's "analyze" button. Stay clear while the machine checks the person's heart rhythm.

If a shock is needed, the AED will let you know when to deliver it. Stand clear of the person and make sure others are clear before you push the AED's "shock" button.

Start or resume CPR until emergency medical help arrives or until the person begins to move. Stay with the person until medical help arrives, and report all of the information you know about what has happened.

DURING THESE CRITICAL MOMENTS, IT IS IMPORTANT TO FOLLOW INSTRUCTIONS GIVEN BY THE AED. THE AED IS PROGRAMMED TO REPEAT THE FOLLOWING CYCLE EVERY TWO MINUTES:

- Evaluation of the person's condition
- Determine if a shock is required
- If no shock required, then complete five cycles of CPR

This cycle should be repeated until you are relieved by a competent person. Remember that an AED will not provide shocks unless it can sense some electrical activity in the heart.

External Breathing Emergencies: Choking

The Conscious Adult

If the person is conscious but not able to breathe or talk, remove any obvious obstruction from the mouth.

- Give back blows.
  - Stand slightly behind the person to one side. Support their chest with one hand and lean the person forward. Give up to five sharp blows between the shoulder blades with the heel of your hand. Stop after each blow to check if the blockage has cleared.

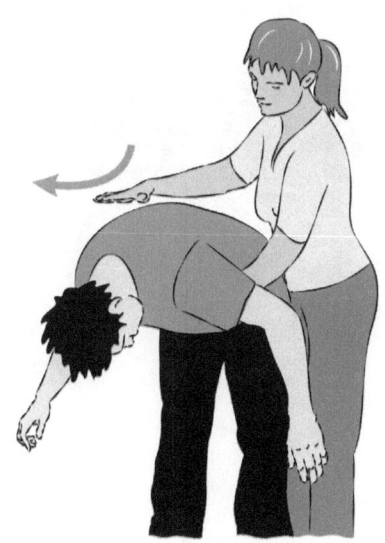

- If the person is still choking, do abdominal thrusts known as the Heimlich maneuver.
  - Stand behind the person, wrap your arms around the waist and bend them well forward.
  - Place your clenched fist just above the person's navel (belly button). Grab your fist with your other hand. Quickly thrust both hands inward and upward.
  - Continue cycles of five back blows and five abdominal thrusts until the object is coughed up or the person starts to breathe or cough.

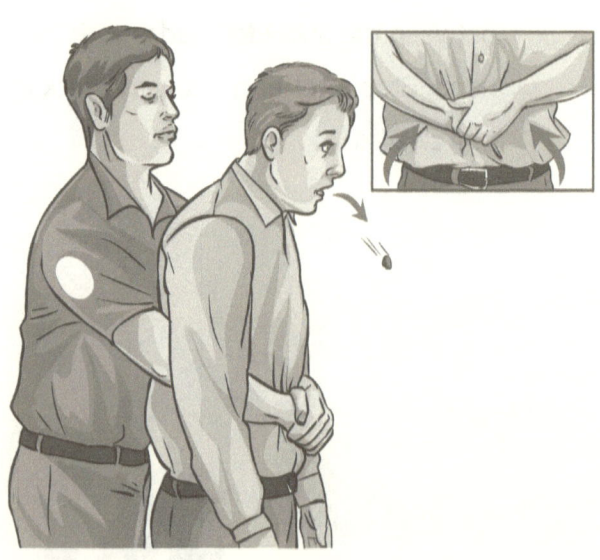

Refrain from the Heimlich maneuver if the person is pregnant, obese, or under one year old. Take the object out of his mouth only if you can see it. Never do a finger sweep unless you can see the object in the person's mouth.

IF THE PERSON IS OBESE OR PREGNANT, DO HIGH ABDOMINAL THRUSTS:

- Stand behind the person, wrap your arms around them and position your hands at the base of the breast bone.
- Quickly pull inward and upward.
- Repeat until the object is dislodged.

The Unconscious Choking Adult

If the person becomes unconscious call 9-1-1 immediately and start CPR.

First Aid

Some form of first aid has existed in human society for thousands of years. First aid not only is a complicated subject in and of itself, but it has a very complicated history. Although we have little information about prehistoric man, they must have been confronted by many situations requiring first aid. For example, they must have developed ways to stop bleeding, to stabilize broken bones or to determine whether a particular plant was poisonous or not.

Over time, certain individuals become more skillful and knowledgeable about how to deal with medical situations. These might have been the first shamans and witchdoctors. Perhaps this was also the beginning of the distinction between medical care which could be provided by the "amateur" or layperson versus the "professional." The distinction continued to develop as medical education and training became more formalized. In time, priests became physicians and barbers became the surgeons.

Another aspect of the history of first aid involved warfare. Men were injured in battles and the lack of medical attention usually resulted in loss of life. In 1099, religious knights trained in medical care organized the Order of St. John to specifically treat battlefield injuries. In other words, although these knights were considered laypersons, they were formally trained to provide "first aid."

It was not until the mid-19th century that the First International Geneva Convention was held, and the Red Cross was created to provide "aid to sick and wounded soldiers in the field." Soldiers were trained to treat their fellow soldiers before the medics arrived.

A decade later, an army surgeon proposed the idea of training civilians in what he termed "pre-medical treatment." The term "first aid" first appeared in 1878 as a combination of "first treatment" and "National Aid." In Britain, civilian ambulance crews were trained specifically for the railways, mines, and the police.

ITEM	MINIMUM QUANTITY
Absorbent compress, 32 sq. in.	1
Adhesive bandages, 1 in. x 3 in.	16
Adhesive tape, 3/8 in. x 2.5 yd. total	1
Antibiotic treatment, 0.14 fl. Oz. (0.9 g)	6
Antiseptic, 0.14 fl. Oz. (0.5 g) application[2]	10
Burn treatment, 1/32 oz. (0.9 g) application[5]	6
First-aid guide[4]	1
Medical exam gloves	2 pairs
Sterile pads, 3 in. x 3 in.	4

Today, first aid is used to give lay responders the basic training to provide aid to the injured or ill in an emergency. Along with training has come the development of tools to aid the lay person: the first aid kit. Requirements for these kits vary according to country, region, occupation and taste. In all cases, minimum requirements are suggested or required. You should check the standard for kits in your region. It is recommended to have some type of kit at all work places, in your home, RV, boat, or car for use during emergencies. Boaters will be familiar with Coast Guard Standards, which differ from the OSHA ANSI standard.

TABLE 160.041–4(b)—ITEMS FOR FIRST-AID KIT

Item	Number per package	Size of package	No. of packages
Bandage compress—4″	1	Single	5
Bandage compress—2″	4	do	2
Waterproof adhesive compress—1″	16	do	2
Triangular bandage—40″	1	do	3
Eye dressing packet, ⅛ ounce Opthalmic ointment, adhesive strips, cotton pads.	3	do	1
Bandage, gauze, compressed, 2 inches by 6 yards	2	do	1
Tourniquet, forceps, scissors, 12 safety pins	1, 1, 1, and 12, respectively	Double	1
Wire splint	1	Single	1
Ammonia inhalants	10	do	1
Iodine applicators (½ ml swab type)	10	do	1
Aspirin, phenacetin and caffeine compound, 6½ gr tablets, vials of 20	5	Double	1
Sterile petrolatum gauze, 3″×18″	4	Single	3

49

Sudden Illnesses

The largest class of problems the lay responder is likely to encounter is in the group of illnesses called "Sudden Illnesses." A sudden illness is any illness that can occur without warning, including diabetes, stroke, seizures, allergic reactions, breathing difficulty, and poisons.

COMMON SIGNALS OF SUDDEN ILLNESSES:

- Changes in level of consciousness such as feeling light-headed, dizzy, drowsy, confused, or becoming unconscious
- Breathing problems (i.e. trouble breathing or no breathing)
- Signals of a possible heart attack
- Signals of a stroke
- Signals of shock
- Loss of vision or blurred vision
- Sweating
- Persistent abdominal pain or pressure
- Nausea or vomiting
- Diarrhea
- Seizures

Diabetes

Although hyperglycemia and hypoglycemia are different
conditions, their major signals are similar and include:

- Changes in the level of consciou
- Changes in mood
- Irregular breathing
- Feeling or looking ill
- Abnormal skin appearance
- Dizziness and headache
- Confusion

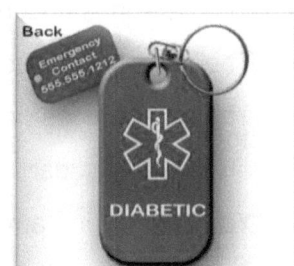

Care for both of these diabetic emergencies is the same.

If you know someone experiencing the signals, you may know the
person is diabetic. A person who is conscious may also tell you he
or she is diabetic. Also look for a medical ID tag, bracelet,
necklace or anklet. Often individuals with diabetes know what is
wrong and will ask for something with sugar in it or they may
carry some form of sugar with them in case they need it. If the
person is conscious and able to swallow and advises you that he
or she needs sugar, give sugar in the form of several glucose
tablets or glucose paste, a 12-ounce serving of fruit juice, milk, or
non-diet soft drink, or table sugar dissolved in a glass of water. If
the problem is too much sugar, this amount of sugar will not
cause further harm.

People who take insulin to control diabetes may have injectable medication with them to care for hyperglycemia. Do *not* try to assist a person by administering insulin to them.

Always call 9-1-1 or the local emergency number if:

- The person is unconscious or about to lose consciousness.
- The person is conscious but unable to swallow.
- The person does not feel better approximately 5 minutes after taking sugar.
- You cannot find any form of sugar immediately.

IF THE PERSON IS UNCONSCIOUS:

- Do not give him or her anything by mouth.
- Give care in the same way you would for any unconscious person.

Stroke

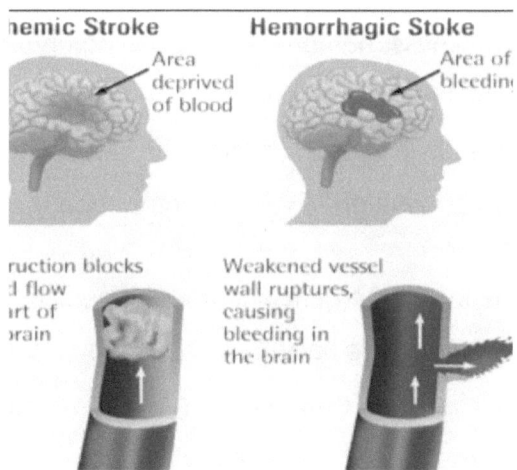

A stroke, also called a cerebrovascular accident (CVA) or brain attack, is caused when blood flow to the brain is cut off or when there is bleeding into the brain. The risk factors for stroke are similar to those for heart disease. Some risk factors are beyond your control such as age, gender, family history of stroke or heart disease.

According to the National Stroke Association, in the United States:

- Nearly 800,000 people experience a new or recurrent stroke every year.
- A stroke happens every 40 seconds.
- Stroke is the fifth leading cause of death in the U.S.
- Every 4 minutes someone dies from stroke.
- Up to 80 percent of strokes can be prevented.
- Stroke is the leading cause of adult disability in the U.S.

Think of the word "FAST" to look for the signals of a stroke in an individual.

- **F**ace: weakness, numbness, or drooping on one side
- **A**rm: weakness or numbness in one arm.
- **S**peech: slurred speech or difficulty speaking.
- **T**ime: determination of when signals began.

Call 9-1-1 or the local emergency number immediately if you encounter someone who is having or has had a stroke, or if the person had a mini-stroke (even if the signals have gone away).

Seizures

If you see someone having a tonic-clonic seizure

Try to lay the person on the floor and gently turn him onto his side.

Move objects in the area out of the way to prevent injury.

Do **NOT** try to stop the person's movements.

Time the seizure. If it lasts longer than 5 minutes or is followed by another seizure, seek emergency treatment.

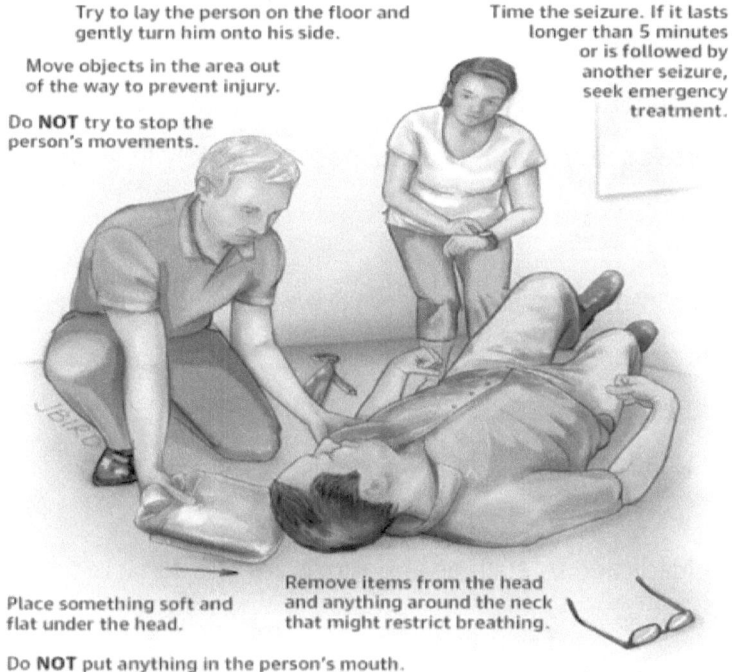

Place something soft and flat under the head.

Remove items from the head and anything around the neck that might restrict breathing.

Do **NOT** put anything in the person's mouth.

Seizures are changes in the brain's electrical activity. This can cause dramatic, noticeable symptoms or even no symptoms at all. The symptoms of a severe seizure are often widely recognized, including violent shaking and loss of control.

However, mild seizures can also be a sign of a significant medical problem, so recognizing them is important. Because some seizures can lead to injury or be evidence of an underlying medical condition, it is important to seek treatment if the person is experiencing them.

Experts classify seizures into two general categories and many subtypes based on the pattern of the attack. Generalized seizures involve both sides of the brain from the start of the attack. Common subtypes include tonic-clonic (grand mal) and absence seizures (petit mal). Febrile and infantile spasms are two types of generalized seizures that occur almost exclusively in young children.

Often the cause of a seizure is unknown. Many conditions can provoke seizures:

- Stroke
- Head injuries
- Very low blood sugar
- Medications, such as antipsychotics and some asthma drugs
- Withdrawal from medications
- Use of drugs such as cocaine and heroin
- Cancer
- Brain infections, such as meningitis

HOW IS THE CONDITION TREATED?

The area around a person should be cleared during a seizure to prevent possible injury. The person should be placed on his or her side with the head cushioned. Stay with the person and contact emergency responders as soon as possible if the seizure lasts longer than two to five minutes, if the person does not awaken after the seizure, or if he or she experiences repeat seizures.

Allergic Reactions

An allergic reaction is the body's way of responding to an "invader." When the body senses a foreign substance, called an antigen, the immune system is triggered. The immune system normally protects the body from harmful agents such as bacteria and toxins. Its overreaction to a harmless substance (an allergen) is called a hypersensitivity reaction, or an allergic reaction.

Most allergic reactions are minor, such as a rash from poison ivy, mosquito or other bug bites, or sneezing from hay fever. The type of reaction depends on the person's immune system response, which is sometimes unpredictable.

In rare cases, an allergic reaction can be life-threatening (known as anaphylaxis). The Asthma and Allergy Foundation of America (AAFA) estimates that at least one in 50 Americans (1.6%), and as many as one in 20 (5.1%) have had anaphylaxis occur, resulting in an average of 186 to 225 deaths per year.

Many people will have some problem with allergies or allergic reactions at some point in their lives. Allergic reactions can range from mild and annoying to sudden and life-threatening. Most allergic reactions are mild, and home treatment can relieve many of the symptoms. An allergic reaction is more serious when severe allergic reaction occurs, when allergies cause other problems (such as nosebleeds, ear problems, wheezing, or coughing), or when home treatment doesn't help.

Call 9-1-1 or activate emergency medical services if someone has any of the following with an allergic reaction:

- Sudden, severe, or rapidly worsening symptoms
- The person tells you they are having a severe reaction
- Exposure to an allergen that previously caused severe or bad reactions
- Swelling of the lips, tongue, or throat
- Wheezing, chest tightness, loud breathing, trouble breathing, or hoarseness of voice
- Confusion, sweating, nausea, or vomiting
- Widespread rash or severe hives
- Lightheadedness, collapse, or unconsciousness

People with known allergies often have emergency medications with them such as an EpiPen, which injects the drug epinephrine. Epinephrine opens the airways and raises blood pressure. This is called a rescue drug. If the person is unable to administer the medication, help him or her to take it.

Breathing Emergencies

There are many different causes for breathing problems. Common causes include:

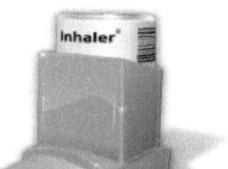

- Asthma
- Chronic obstructive pulmonary disease (COPD), sometimes called emphysema or chronic bronchitis
- Life-threatening allergic reaction
- Choking

If someone is having breathing difficulty, immediately call 9-1-1 or your local emergency number, then:

- Check the person's airway, breathing, and pulse. If necessary, begin CPR.
- Loosen any tight clothing.
- Help the person use any prescribed medication (an asthma inhaler or home oxygen).
- Continue to monitor the person's breathing and pulse until medical help arrives. DO NOT assume that the person's condition is improving if you can no longer hear abnormal breath sounds like wheezing.

Poisoning

If you suspect a poisoning:

- Check the scene and the person. Try to find out what poison was taken.
- Call the National Poison Control Center at (800) 222-1222 and follow their instructions, then call 9-1-1.
- Care for any life-threatening conditions found.

Soft Tissue Injuries (Wounds)

Soft tissue injuries are injuries causing damage to the skin, muscles, ligaments, or tendons of the body and fall into two main groups: wounds and the special category of burns.

TYPES OF WOUNDS

- Lacerations – Injury where tissue is cut or torn
- Abrasions – Injury where a superficial layer of tissue is removed, as seen with 1st degree burns.
- Punctures – Injuries resulting from a narrow penetration of the soft tissue.
- Avulsions – Injuries where a section of tissue is torn off, either partially or in total. In partial avulsions, the tissue is elevated but remains attached to the body. A total avulsion means that the tissue is completely torn from the body with no point of attachment.

TREATMENT FOR MOST WOUNDS LESS THAN TWO (2) INCHES IN LENGTH IS AS FOLLOWS:

- Wash and disinfect the wound to removing all dirt and debris.
- Use direct pressure and elevation to control bleeding and swelling.
- Bandage the wound with a sterile dressing or bandage (very minor wounds may heal fine without a bandage).
- Keep the wound clean and dry.
- Consult a Doctor.

Bleeding will occur with most wounds. Control of bleeding can be accomplished using direct pressure in ninety percent of injuries. The procedure for controlling bleeding is as follows:

- After taking infection control precautions for the caregiver, direct pressure is applied to the injury with sterile gauze. If bleeding is profuse or seeps through the gauze, add more gauze, but do not remove the existing pieces. This will prevent the clotting process from being interrupted.
- If bleeding continues to be severe, the extremity or body part should be elevated above the level of the heart. This will decrease the amount of blood flowing to the injury site by using gravity to help decrease the amount of blood flow.

- If bleeding continues, add more gauze to the existing dressing and tie a pressure bandage to the site of injury.
- If direct pressure, elevation, and pressure bandage fail to control bleeding, apply pressure to a pressure point of the injury if it is to an extremity. This will aid in further decreasing the flow of blood to the injury site.
- A tourniquet should not be used unless the bleeding is so bad the person will die if it is not controlled. Tourniquets are to be used in rare circumstances and only by trained healthcare professionals. Once a tourniquet is applied, it is not to be removed unless so directed by a physician.

In cases where the bleeding cannot be controlled with direct pressure, the use of a tourniquet or indirect pressure may be required. If you are using a commercial tourniquet, follow the manufacturer's instructions. If a commercial tourniquet is not available, one can be created by using a triangular bandage, belt or anything that is two inches wide. Never use anything that is less than two inches in width. Tourniquets should only be used on limbs (arms and legs).

Apply the tourniquet by wrapping it between the wound and the heart, approximately two inches above the wound.

Using a stick or similar object, tie a half knot and insert the stick as shown in the picture.

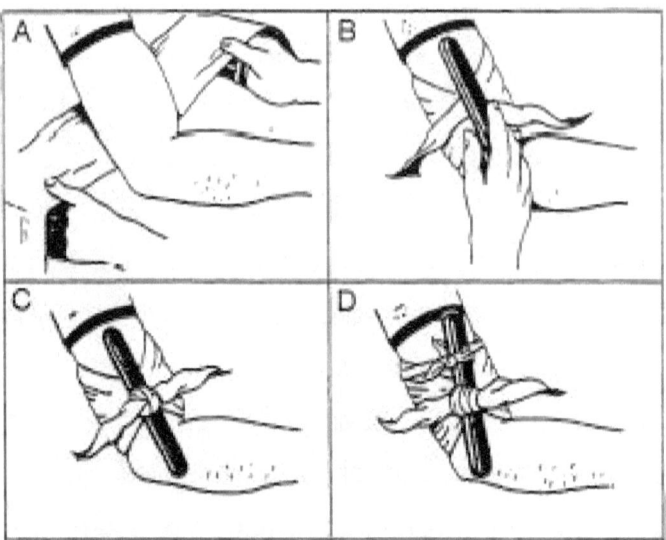

- Make sure you draw the tourniquet tight enough to stop the bleeding but do not make it tighter than necessary.
- Never loosen a tourniquet once it has been applied. The loosening of a tourniquet may dislodge clots and result enough in blood loss to cause severe shock and death.

Managing Shock

Shock may result from trauma, heatstroke, blood loss, an allergic reaction, severe infection, poisoning, severe burns or other causes. When a person is in shock, his or her organs aren't getting enough blood or oxygen. If untreated, this can lead to permanent organ damage or even death.

THE MAIN TYPES OF SHOCK INCLUDE:

- Cardiogenic shock (due to heart problems)
- Hypovolemic shock (caused by too little blood volume)
- Anaphylactic shock (caused by allergic reaction)
- Septic shock (due to infections)
- Neurogenic shock (caused by damage to the nervous system)

SIGNS AND SYMPTOMS OF SHOCK VARY DEPENDING ON CIRCUMSTANCES AND MAY INCLUDE:

- Cool, clammy skin
- Pale or ashen skin
- Rapid pulse
- Rapid breathing
- Nausea or vomiting
- Enlarged pupils
- Weakness or fatigue
- Dizziness or fainting
- Changes in mental status or behavior (i.e. anxiousness or agitation)

IF YOU SUSPECT A PERSON IS IN SHOCK, *CALL 9-1-1 OR YOUR LOCAL EMERGENCY NUMBER*. THEN IMMEDIATELY TAKE THE FOLLOWING STEPS:

- Lay the person down and elevate the legs and feet slightly, unless you think this may cause pain or further injury.
- Keep the person still and don't move him or her unless necessary.
- Begin CPR if the person shows no signs of life such as breathing, coughing or movement.
- Loosen tight clothing and, if needed, cover the person with a blanket to prevent chilling.
- Don't let the person eat or drink anything.
- If the person vomits or begins bleeding from the mouth, turn him or her onto a side to prevent choking, unless you suspect a spinal injury.

Soft Tissue injuries (Burns)

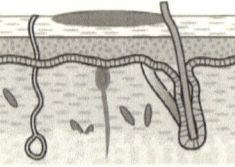
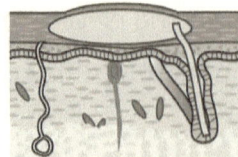
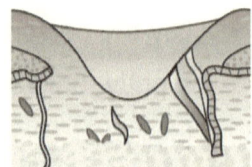
FOR MINOR BURNS:

- Cool the burn to help soothe the pain. Hold the burned area under cool (not cold) running water for 10 to 15 minutes or until the pain eases. Or apply a clean towel dampened with cool tap water.

- Remove rings or other tight items from the burned area. Try to do this quickly and gently, before the area swells.

- Don't break small blisters (no bigger than your little fingernail). If blisters break, gently clean the area with mild soap and water, apply an antibiotic ointment, and cover it with a nonstick gauze bandage.

See your doctor if you develop large blisters. Large blisters are best removed, as they rarely will remain intact on their own. Also seek medical help if the burn covers a large area of the body or if you notice signs of infection, such as oozing from the wound and increased pain, redness and swelling.

FOR LARGE BURNS:

Call 9-1-1 or emergency medical help for major burns. Until an emergency unit arrives, take these actions:

- Protect the burned person from further harm. If you can do so safely, make sure the person you're helping is not in contact with smoldering materials or exposed to smoke or heat. But don't remove burned clothing stuck to the skin.
- Check for signs of circulation. Look for breathing, coughing or movement. Begin CPR if needed.
- Remove jewelry, belts and other restrictive items, especially from around burned areas and the neck. Burned areas swell rapidly.
- Don't immerse large severe burns in cold water. Doing so could cause a serious loss of body heat (hypothermia) or a drop in blood pressure and decreased blood flow (shock).
- Elevate the burned area. Raise the wound above heart level, if possible.
- Cover the area of the burn. Use a cool, moist, bandage or a clean cloth.

Environmental Illnesses

Normal body temperature varies by person, age, activity, and time of day. The average normal body temperature

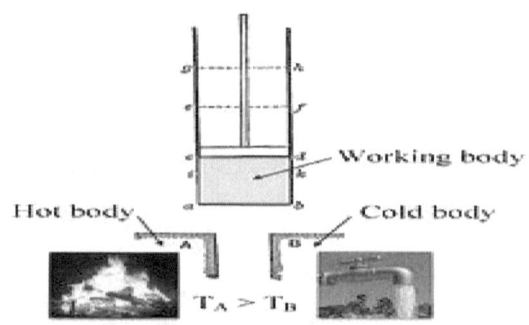

Working body

Hot body

Cold body

$T_A > T_B$

is generally accepted as 98.6°F (37°C). Some studies have shown that the "normal" body temperature can have a wide range, from 97°F (36.1°C) to 99°F (37.2°C).

When the body gets too hot, it uses several strategies to cool down, including sweating. But if a person spends too much time in the heat without taking in enough fluids, the body's cooling processes can't work properly. This condition is known as hyperthermia and can result in the following:

Heat cramps: painful muscle spasms in the abdomen, arms or legs following strenuous activity. Heat cramps are caused by a lack of salt in the body.

- For heat cramps, give the person liquid and get them out of the heat.

Heat exhaustion: a warning that the body is getting too hot. The person may be thirsty, giddy, weak, uncoordinated, nauseated and sweating profusely. The body temperature is normal and the pulse is normal or raised. The skin is cold and clammy.

- For heat exhaustion, control liquid intake to four-ounce increments, if the person can manage it, and remove from the heat to shelter.

Heat stroke can be life-threatening and victims can die. A person with heat stroke usually has a body temperature above 104 degrees Fahrenheit. Other symptoms include confusion, combativeness, bizarre behavior, faintness, staggering, strong and rapid pulse, and possible delirium or coma. High body temperature is capable of producing irreversible brain damage.

- The first step in treating heat stroke is to reduce body temperature by cooling the body from the outside. This can be done by removing tight or unnecessary clothing, spraying the person with water, blowing cool air on the person, or wrapping the person loosely in wet sheets. Alternatively, ice packs can be placed at the neck, groin and armpits to accelerate cooling. Medical personnel should be called immediately.

Hyponatremia

Hyponatremia occurs when the combination of high water intake and low renal output produces a diluted blood serum sodium. This often occurs when persons working in hot weather consume large amounts of water without taking in additional sodium.

Hyponatremia can have serious consequences, including brain disease, cardiac arrest, cerebral edema, seizures, coma, and death.

The warning signs of hyponatremia look a lot like the symptoms of heatstroke and exhaustion. The person might be hot, have a headache, and just feel crummy. Other early symptoms can include diarrhea, nausea, and vomiting.

If you see someone with these symptoms, pull them aside, put them in the shade and talk to them, but give them no water.

Hypothermia

Hypothermia is a medical emergency that occurs when the body loses heat faster than it can produce heat, causing a dangerously low body temperature. Normal body temperature is around 98.6 F (37 C). Hypothermia occurs as body temperature passes below 95 F (35 C).

When body temperature drops, the heart, nervous system and other organs can't work normally. Left untreated, hypothermia can eventually lead to complete failure of the heart and respiratory system and death.

Hypothermia is most often caused by exposure to cold weather or immersion in a cold body of water. Primary treatments for hypothermia are methods to warm the body back to a normal temperature.

Frostbite

Frostbite is an injury to the body that is caused by freezing. It most often affects the nose, ears, cheeks, chin, fingers, or toes. Frostbite can permanently damage the body, and severe cases can lead to amputation.

Signs of frostbite include:
- A white or grayish-yellow skin area
- Skin that feels unusually firm or waxy
- Numbness

If you have symptoms of frostbite, seek medical care. But if immediate medical care isn't available, here are steps to take:
- Get into a warm room as soon as possible.
- Unless absolutely necessary, do not walk on frostbitten feet or toes. Walking increases the damage.
- Put the affected area in warm – not hot – water.
- You can also warm the affected area using body heat. For example, use your armpit to warm frostbitten fingers.
- Don't rub the frostbitten area with snow or massage it at all. This can cause more damage.
- Don't use a heating pad, heat lamp, or the heat of a stove, fireplace, or radiator for warming. Since frostbite makes an area numb, you could burn it.

Bone and Joint Injuries

The most common types of injuries.

The most common leg and arm injuries are fractures, sprains, strains, and bruises. Some injuries can be treated at home, while others need to be treated or checked by a physician.

FRACTURES:

A fracture is a break or crack in the bone and needs to be treated by a doctor. If you think your child has a broken bone, follow the first aid instructions below.

- Shoulder or arm: Use a sling made of a triangular piece of cloth. A cold pack may help. Drive your child to the doctor.
- Leg: Use padded boards, pillows, or newspapers to splint the fracture. At a minimum, carry your child and don't permit your child to put any weight on the leg. A cold pack may help. Drive your child to the doctor.
- Neck: Protect the neck from any turning or bending. Do not move your child until a neck brace or spine board has been applied. Call a rescue squad (9-1-1) for help.

SPRAINS:

Sprains are stretches or tears of ligaments (bands of tissue that connect one bone to another). They are caused by sudden twisting injuries and require medical attention (unless they are very mild). Knees and ankles are often sprained.

- Immediately wrap the injured area with an elastic bandage and put ice on the injury to reduce bleeding, swelling, and pain.
- While some mild sprains can be cared for at home, most injuries to ligaments need to be checked by your healthcare provider. You can drive your child to the doctor.

Treat with R.I.C.E. (rest, ice, compression, and elevation) for the first 24 to 48 hours.

- Apply compression with a snug, elastic bandage for 48 hours. Numbness, tingling, or increased pain means the bandage is too tight.
- Apply a cold pack or crushed ice in a plastic bag for 20 minutes. Avoid frostbite. Repeat every hour for 4 hours.
- Give acetaminophen or ibuprofen as needed for pain relief. Continue for at least 48 hours.
- Keep injured ankle or knee elevated and at rest for 24 hours. After 24 hours, allow any activity that doesn't cause pain.
- After 48 hours, you can use a heating pad for 10 minutes a few times per day to help absorb the blood.

CERVICAL INJURY:

Never move anyone who you think may have a spinal injury, unless it is absolutely necessary. For example, if you need to get the person out of a burning car or help them to breathe.

Keep the person absolutely still and safe until medical help arrives.
- Call 9-1-1.
- Hold the person's head and neck in the position in which they were found. Do not try to straighten the neck. Do not allow the neck to bend or twist.
- Do not allow the person to get up and walk unassisted.

If the person is not alert or responding:
- Check the person's breathing and circulation. If necessary, begin rescue breathing and CPR.
- Do not tilt the head back when doing CPR. Do not do rescue breathing, do chest compressions only.

Do not roll the person over unless the person is vomiting or choking on blood, or you need to check for breathing. If you need to roll the person over:
- Have someone assist you.
- One person should be located at the person's head, the other at the person's side.
- Keep the person's head, neck, and back in line while you roll him or her onto one side.

CRANIAL INJURIES:

Most head trauma involves injuries that are minor and don't require specialized attention or hospitalization. However, even minor injuries may cause persistent chronic symptoms such as headache or difficulty concentrating.

Call 9-1-1 or your local emergency number if any of the following signs or symptoms are apparent, because they may indicate a more serious head injury:

- Severe head or facial bleeding
- Bleeding or fluid leakage from the nose or ears
- Severe headache
- Change in level of consciousness
- Black-and-blue discoloration below the eyes or behind the ears
- Cessation of breathing
- Confusion
- Loss of balance
- Weakness or an inability to use an arm or leg
- Unequal pupil size
- Slurred speech
- Seizures

References

The Life Skills Training Manual of Arthur Jackson CTC INC. is based on the following materials and other sources as cited in the text;

- *2015 International Consensus on Cardiopulmonary Resuscitation and Emergency Cardiovascular Care Science with Treatment Recommendations*

- *2005 American Heart Association guidelines for cardiopulmonary resuscitation and emergency cardiovascular care part 14: first aid. Circulation. 2005; 112(suppl):IV-196 –IV-203*

- *2010 American Heart Association and American Red Cross Guidelines for First Aid, Circulation.2010; 122: S934-S946 doi: 10.1161/CIRCULATIONAHA.110.971150*

- *Best Practices Guide: Fundamentals of a Workplace First-Aid Program (PDF *). OSHA Publication 3317-06N, (2006). Safety and Health Topics | Medical and First Aid* https://www.osha.gov/SLTC/medicalfirstaid/index.html

- *First aid burns, http://www.mayoclinic.org/first-aid/first-aid-burns/basics/art-20056649, accessed 8/1/2015*

- *Burns. The Merck Manual Professional Edition. http://www.merckmanuals.com/professional/injuries_poiso ning/burns/burns.html?qt=burns&alt=sh#v1112914. Accessed Oct. 23, 2014.*
- *What to do in a medical emergency. American College of Emergency Physicians. http://www.emergencycareforyou.org. Accessed 7/14/2015.*

INDEX

Partial listing of online classes offered by Arthur Jackson CTC by country;

Udemy University classes offered in United States, Canada, South America, Asia, and Africa

Inequality in the modern world
"Arthur Jackson, CEO Arthur Jackson CTC, Professor, Engineer, Attorney "
Free • ★★★★☆ 4 Reviews • 7 lectures, 37 mins video • All Levels

Disaster Planning
"Arthur Jackson, CEO Arthur Jackson CTC, Professor, Engineer, Attorney "
Free • ★★★★☆ 4 Reviews • 9 lectures, 33 mins video • Beginner Level

Update on evolving Techniques in First Aid
"Arthur Jackson, CEO Arthur Jackson CTC, Professor, Engineer, Attorney "
Free • ★★★★★ 7 Reviews • 6 lectures, 37 mins video • All Levels

The Emergency Medical System
"Arthur Jackson, CEO Arthur Jackson CTC, Professor, Engineer, Attorney "
Free • ★★★★★ 5 Reviews • 9 lectures, 36 mins video • All Levels

First Aid
"Arthur Jackson, CEO Arthur Jackson CTC, Professor, Engineer, Attorney "
Free • ★★★★☆ 43 Reviews • 10 lectures, 1 hour video • Beginner Level

RESILIENCE: Returning to normal after an emergency
"Arthur Jackson, CEO Arthur Jackson CTC, Professor, Engineer, Attorney "
Free • ★★★★★ 6 Reviews • 7 lectures, 41 mins video • Beginner Level

Climate and catastrophe and war
"Arthur Jackson, CEO Arthur Jackson CTC, Professor, Engineer, Attorney "
$12 • ★★★★★ No reviews yet • 7 lectures, 33 mins video • All Levels

Cardio Pulmonary Resuscitation (CPR)
"Arthur Jackson, CEO Arthur Jackson CTC, Professor, Engineer, Attorney "
Free • ★★★★☆ 12 Reviews • 7 lectures, 44 mins video • Beginner Level

Introduction to Emergencies for Lay Responders
"Arthur Jackson, CEO Arthur Jackson CTC, Professor, Engineer, Attorney "
Free • ★★★★★ 5 Reviews • 8 lectures, 1 hour video • Beginner Level

First Aid for Remote and Rural Locations
"Arthur Jackson, CEO Arthur Jackson CTC, Professor, Engineer, Attorney "

Free • ★ ★ ★ ★ ★ 6 Reviews • 25 lectures, 2 hours video • Expert Level

Disease Transmission Control for the Lay Person
"Arthur Jackson, CEO Arthur Jackson CTC, Professor, Engineer, Attorney "

Free • ★ ★ ★ ★ ★ 7 Reviews • 9 lectures, 32 mins video • Beginner Level

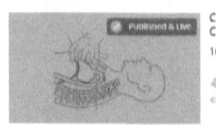
Published & Live

Cardio Pulmonary Resuscitation-CPR

16 Learners ★ ★ ★ ★ ★

FREE
Views 447

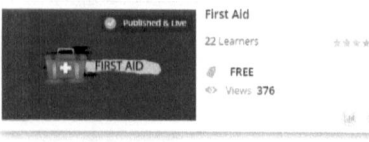
Published & Live

First Aid

22 Learners ★ ★ ★ ★ ★

FREE
Views 376

Published & Live

Introduction to Emergencies for Lay responders

2 Learners ★ ★ ★ ★ ★

FREE
Views 108

Published & Live

Disaster Planning

8 Learners ★ ★ ★ ★ ★

FREE
Views 213

Class Evaluation Sheet

Please provide your feedback on the course instruction and content. This helps maintain the high quality and standards we seek and you deserve.

Class Information:

Instructor: _____ Date of class_____

Class title:

_____

Use this key to rank the items below;

5-Excellent, **4**-Very Good, **3**- Good, **2**- Fair, **1**- Poor

Class Section   Rating

	5	4	3	2	1
Responder					
First Aid					
CPR					
Choking					
Sudden illness					
Soft tissue					
Burns					
Bones and joints					

Did you watch/complete any online course prior to taking this course? Yes_____ No_____

Was the class conducted in a professional manner? Yes___ No__

Comments:

_____

_____

_____

_____

www.ingramcontent.com/pod-product-compliance
Lightning Source LLC
Chambersburg PA
CBHW022123170526
45157CB00004B/1731